Kepler

by

Walter W. Bryant

Kepler
by Walter W. Bryant

ISBN: 978-93-59320-24-3

Published by

DOUBLE 9 BOOKS

2/13-B, Ansari Road
Daryaganj, New Delhi – 110002
info@double9books.com
www.double9books.com
Tel. 011-40042856

ABOUT THE AUTHOR

Walter W. Bryant was a skilled author recognized for his profound literary contributions, particularly "Kepler." He was born and died in 1865 to 1923. His writing assists people in connecting with and understanding one another. The author's works are full of originality and passion, transporting readers to other realms and emotions. Bryant's works stretched the boundaries of conventional storytelling, providing readers with a unique blend of science, philosophy, and imaginative thinking. He possessed a sharp mind and an in-depth understanding of science and the human circumstance. Bryant's literary career, which began in 1865, was marked by a keen interest in the natural world and cosmic mysteries. "Kepler," one of his most notable works, illustrates his love with science as well as his ability to merge scientific principles into interesting fiction. Bryant probably wrote about the life and achievements of Johannes Kepler, the famed German astronomer known for his laws of planetary motion, in "Kepler." While information on Walter W. Bryant's life and works is limited, his commitment to combining literature and science in "Kepler" indicates his desire to expand the scope of literary inquiry.

CONTENTS

Chapter I
Astronomy Before Kepler

In order to emphasise the importance of the reforms introduced into astronomy by Kepler, it will be well to sketch briefly the history of the theories which he had to overthrow. In very early times it must have been realised that the sun and moon were continually changing their places among the stars. The day, the month, and the year were obvious divisions of time, and longer periods were suggested by the tabulation of eclipses. We can imagine the respect accorded to the Chaldaean sages who first discovered that eclipses could be predicted, and how the philosophers of Mesopotamia must have sought eagerly for evidence of fresh periodic laws. Certain of the stars, which appeared to wander, and were hence called planets, provided an extended field for these speculations. Among the Chaldaeans and Babylonians the knowledge gradually acquired was probably confined to the priests and utilised mainly for astrological prediction or the fixing of religious observances. Such speculations as were current among them, and also among the Egyptians and others who came to share their knowledge, were almost entirely devoted to mythology, assigning fanciful terrestrial origins to constellations, with occasional controversies as to how the earth is supported in space. The Greeks, too, had an elaborate mythology largely adapted from their neighbours, but they were not satisfied with this, and made persistent attempts to reduce the apparent motions of celestial objects to geometrical laws. Some of the Pythagoreans, if not Pythagoras himself, held that the earth is a sphere, and that the apparent daily revolution of the sun and stars is really due to a motion of the earth, though at first this motion of the earth was not supposed to be one of rotation about an axis. These notions, and also that the planets on the whole move round from west to east with reference to the stars, were made known to a larger circle through the writings of Plato. To Plato moreover is attributed the challenge to astronomers to represent all the motions of the heavenly bodies by uniformly described circles, a challenge generally held responsible for a vast amount of wasted effort, and the postponement, for many centuries, of real progress. Eudoxus of Cnidus, endeavouring to account for the fact that the planets, during every apparent revolution round the earth, come to rest twice, and in the shorter interval between these "stationary points," move in the opposite direction, found that he could represent the phenomena fairly well by a system of concentric spheres, each rotating with its own velocity, and carrying its own particular planet round its own equator, the outermost sphere carrying the fixed stars. It was necessary to assume that the axes about which the various spheres revolved should have circular motions also, and gradually an increased number of spheres was evolved, the total number required by Aristotle reaching fifty-five. It may be regarded as counting in Aristotle's favour that he did consider the earth to be a sphere and not a flat disc, but he seems to have thought that the mathematical spheres of Eudoxus had a real solid existence, and that not only meteors, shooting stars and aurora, but also comets and the milky way belong to the atmosphere. His really great service to science in collating and criticising all that was known of natural science would have been greater if so much of the discussion had not been on the exact meaning of words used to describe phenomena, instead of on the facts and causes of the phenomena themselves.

Aristarchus of Samos seems to have been the first to suggest that the planets revolved not about the earth but about the sun, but the idea seemed so improbable that it was hardly noticed, especially as Aristarchus himself did not expand it into a treatise.

About this time the necessity for more accurate places of the sun and moon, and the liberality of the Ptolemys who ruled Egypt, combined to provide regular observations at Alexandria, so that, when Hipparchus came upon the scene, there was a considerable amount of material for him to use. His discoveries marked a great advance in the science of astronomy. He noted the irregular motion of the sun, and, to explain it, assumed that it revolved uniformly not exactly about the earth but about a point some distance away, called the "excentric".[1] The line joining the centre of the earth to the excentric passes through the apses of the sun's orbit, where its distance from the earth is greatest and least. The same result he could obtain by assuming that the sun moved round a small circle, whose centre described a larger circle about the earth; this larger circle carrying the other was called the "deferent": so that the actual motion of the sun was in an epicycle. Of the two methods of expression Hipparchus ultimately preferred the second. He applied the same process to the moon but found that he could depend upon its being right only at new and full moon. The irregularity at first and third quarters he left to be investigated by his successors. He also considered the planetary observations at his disposal insufficient and so gave up the attempt at a complete planetary theory. He made improved determinations of some of the elements of the motions of the sun and moon, and discovered the Precession of the Equinoxes, from the Alexandrian observations which showed that each year as the sun came to cross the equator at the vernal equinox it did so at a point about fifty seconds of arc earlier on the ecliptic, thus producing in 150 years an unmistakable change of a couple of degrees, or four times the sun's diameter. He also invented trigonometry. His star catalogue was due to the appearance of a new star which caused him to search for possible previous similar phenomena, and also to prepare for checking future ones. No advance was made in theoretical astronomy for 260 years, the interval between Hipparchus and Ptolemy of Alexandria. Ptolemy accepted the spherical form of the earth but denied its rotation or any other movement. He made no advance on Hipparchus in regard to the sun, though the lapse of time had largely increased the errors of the elements adopted by the latter. In the case of the moon, however, Ptolemy traced the variable inequality noticed sometimes by Hipparchus at first and last quarter, which vanished when the moon was in apogee or perigee. This he called the evection, and introduced another epicycle to represent it. In his planetary theory he found that the places given by his adopted excentric did not fit, being one way at apogee and the other at perigee; so that the centre of distance must be nearer the earth. He found it best to assume the centre of distance half-way between the centre of the earth and the excentric, thus "bisecting the excentricity". Even this did not fit in the case of Mercury, and in general the agreement between theory and observation was spoilt by the necessity of making all the orbital planes pass through the centre of the earth, instead of the sun, thus making a good accordance practically impossible.

Footnote 1: See Glossary for this and other technical terms.

After Ptolemy's time very little was heard for many centuries of any fresh planetary theory, though advances in some points of detail were made, notably by some of the Arab philosophers, who obtained improved values for some of the elements by using better instruments. From time to time various modifications of Ptolemy's theory were suggested, but none of any real value. The Moors in Spain did their share of the work carried on by their Eastern co-religionists, and the first independent star catalogue since the time of Hipparchus was made by another Oriental, Tamerlane's grandson, Ulugh Begh, who built a fine observatory at Samarcand in the fifteenth century. In Spain the work was not monopolised by the Moors, for in the thirteenth century Alphonso of Castile, with the assistance of Jewish and Christian computers, compiled the Alphonsine tables, completed in 1252, in which year he ascended the throne as Alphonso X. They were long circulated in MS. and were first printed in 1483, not long before the end of the period of stagnation.

Copernicus was born in 1473 at Thorn in Polish Prussia. In the course of his studies at Cracow and at several Italian universities, he learnt all that was known of the Ptolemaic astronomy and determined to reform it. His maternal uncle, the Bishop of Ermland, having provided him with a lay canonry in the Cathedral of Frauenburg, he had leisure to devote himself to Science. Reviewing the suggestions of the ancient Greeks, he was struck by the simplification that would be introduced by reviving the idea that the annual motion should be attributed to the earth itself instead of having a separate annual epicycle for each planet and for the sun. Of the seventy odd circles or epicycles required by the latest form of the Ptolemaic system, Copernicus succeeded in dispensing with rather more than half, but he still required thirty-four, which was the exact number assumed before the time of Aristotle. His considerations were almost entirely mathematical, his only invasion into physics being in defence of the "moving earth" against the stock objection that if the earth moved, loose objects would fly off, and towers fall. He did not break sufficiently away from the old tradition of uniform circular motion. Ptolemy's efforts at exactness were baulked, as we have seen, by the supposed necessity of all the orbit planes passing through the earth, and if Copernicus had simply transferred this responsibility to the sun he would have done better. But he would not sacrifice the old fetish, and so, the orbit of the earth being clearly not circular with respect to the sun, he made all his planetary planes pass through the centre of the earth's orbit, instead of through the sun, thus handicapping himself in the same way though not in the same degree as Ptolemy. His thirty-four circles or epicycles comprised four for the earth, three for the moon, seven for Mercury (on account of his highly eccentric orbit) and five each for the other planets.

It is rather an exaggeration to call the present accepted system the Copernican system, as it is really due to Kepler, half a century after the death of Copernicus, but much credit is due to the latter for his successful attempt to provide a real alternative for the Ptolemaic system, instead of tinkering with it. The old geocentric system once shaken, the way was gradually smoothed for the heliocentric system, which Copernicus, still hampered by tradition, did not quite reach. He was hardly a practical astronomer in the observational sense. His first recorded observation, of an occultation of

Aldebaran, was made in 1497, and he is not known to have made as many as fifty astronomical observations, while, of the few he did make and use, at least one was more than half a degree in error, which would have been intolerable to such an observer as Hipparchus. Copernicus in fact seems to have considered accurate observations unattainable with the instruments at hand. He refused to give any opinion on the projected reform of the calendar, on the ground that the motions of the sun and moon were not known with sufficient accuracy. It is possible that with better data he might have made much more progress. He was in no hurry to publish anything, perhaps on account of possible opposition. Certainly Luther, with his obstinate conviction of the verbal accuracy of the Scriptures, rejected as mere folly the idea of a moving earth, and Melanchthon thought such opinions should be prohibited, but Rheticus, a professor at the Protestant University of Wittenberg and an enthusiastic pupil of Copernicus, urged publication, and undertook to see the work through the press. This, however, he was unable to complete and another Lutheran, Osiander, to whom he entrusted it, wrote a preface, with the apparent intention of disarming opposition, in which he stated that the principles laid down were only abstract hypotheses convenient for purposes of calculation. This unauthorised interpolation may have had its share in postponing the prohibition of the book by the Church of Rome.

According to Copernicus the earth is only a planet like the others, and not even the biggest one, while the sun is the most important body in the system, and the stars probably too far away for any motion of the earth to affect their apparent places. The earth in fact is very small in comparison with the distance of the stars, as evidenced by the fact that an observer anywhere on the earth appears to be in the middle of the universe. He shows that the revolution of the earth will account for the seasons, and for the stationary points and retrograde motions of the planets. He corrects definitely the order of the planets outwards from the sun, a matter which had been in dispute. A notable defect is due to the idea that a body can only revolve about another body or a point, as if rigidly connected with it, so that, in order to keep the earth's axis in a constant direction in space, he has to invent a third motion. His discussion of precession, which he rightly attributes to a slow motion of the earth's axis, is marred by the idea that the precession is variable. With all its defects, partly due to reliance on bad observations, the work showed a great advance in the interpretation of the motions of the planets; and his determinations of the periods both in relation to the earth and to the stars were adopted by Reinhold, Professor of Astronomy at Wittenberg, for the new Prutenic or Prussian Tables, which were to supersede the obsolete Alphonsine Tables of the thirteenth century.

In comparison with the question of the motion of the earth, no other astronomical detail of the time seems to be of much consequence. Comets, such as from time to time appeared, bright enough for naked eye observation, were still regarded as atmospheric phenomena, and their principal interest, as well as that of eclipses and planetary conjunctions, was in relation to astrology. Reform, however, was obviously in the air. The doctrine of Copernicus was destined very soon to divide others besides the Lutheran leaders. The leaven of inquiry was working, and not long after the death of Copernicus real advances were to come, first in the accuracy of observations, and, as a necessary result of these, in the planetary theory itself.

Chapter II
Early Life of Kepler

On 21st December, 1571, at Weil in the Duchy of Wurtemberg, was born a weak and sickly seven-months' child, to whom his parents Henry and Catherine Kepler gave the name of John. Henry Kepler was a petty officer in the service of the reigning Duke, and in 1576 joined the army serving in the Netherlands. His wife followed him, leaving her young son in his grandfather's care at Leonberg, where he barely recovered from a severe attack of smallpox. It was from this place that John derived the Latinised name of Leonmontanus, in accordance with the common practice of the time, but he was not known by it to any great extent. He was sent to school in 1577, but in the following year his father returned to Germany, almost ruined by the absconding of an acquaintance for whom he had become surety. Henry Kepler was obliged to sell his house and most of his belongings, and to keep a tavern at Elmendingen, withdrawing his son from school to help him with the rough work. In 1583 young Kepler was sent to the school at Elmendingen, and in 1584 had another narrow escape from death by a violent illness. In 1586 he was sent, at the charges of the Duke, to the monastic school of Maulbronn; from whence, in accordance with the school regulations, he passed at the end of his first year the examination for the bachelor's degree at Tübingen, returning for two more years as a "veteran" to Maulbronn before being admitted as a resident student at Tübingen. The three years thus spent at Maulbronn were marked by recurrences of several of the diseases from which he had suffered in childhood, and also by family troubles at his home. His father went away after a quarrel with his wife Catherine, and died abroad. Catherine herself, who seems to have been of a very unamiable disposition, next quarrelled with her own relatives. It is not surprising therefore that Kepler after taking his M.A. degree in August, 1591, coming out second in the examination lists, was ready to accept the first appointment offered him, even if it should involve leaving home. This happened to be the lectureship in astronomy at Gratz, the chief town in Styria. Kepler's knowledge of astronomy was limited to the compulsory school course, nor had he as yet any particular leaning towards the science; the post, moreover, was a meagre and unimportant one. On the other hand he had frequently expressed disgust at the way in which one after another of his companions had refused "foreign" appointments which had been arranged for them under the Duke's scheme of education. His tutors also strongly urged him to accept the lectureship, and he had not the usual reluctance to leave home. He therefore proceeded to Gratz, protesting that he did not thereby forfeit his claim to a more promising opening, when such should appear. His astronomical tutor, Maestlin, encouraged him to devote himself to his newly adopted science, and the first result of this advice appeared before very long in Kepler's "Mysterium Cosmographicum". The bent of his mind was towards philosophical speculation, to which he had been attracted in his youthful studies of Scaliger's "Exoteric Exercises". He says he devoted much time "to the examination of the nature of heaven, of souls, of genii, of the elements, of the essence of fire, of the cause of fountains, the ebb and flow of the tides, the shape of the continents and inland seas, and things of this sort". Following his tutor in his admiration for the Copernican theory, he wrote an essay on the primary motion, attributing it to the rotation of the earth, and this not for the mathematical reasons brought forward by Copernicus, but, as he himself says, on physical or metaphysical grounds. In 1595, having more leisure from lectures, he turned his speculative mind to the number, size, and motion of the planetary orbits. He first tried simple numerical relations, but none of them appeared to be twice, thrice, or four times as great as another, although he felt convinced that there was some relation between the motions and the distances, seeing that when a gap appeared in one series, there was a corresponding gap in the other. These gaps he attempted to fill by hypothetical planets between Mars and Jupiter, and between Mercury and Venus, but this method also failed to provide the regular proportion which he sought, besides being open to the objection that on the same principle there might be many more equally invisible planets at either end of the series. He was nevertheless unwilling to adopt the opinion of Rheticus that the number six was sacred, maintaining that the "sacredness" of the number was of much more recent date than the creation of the worlds, and could not therefore account for it. He next tried an ingenious idea, comparing the perpendiculars from different points of a quadrant of a circle on a tangent at its extremity. The greatest of these, the tangent, not being cut by the quadrant, he called the line of the sun, and associated with infinite force. The shortest, being the point at the other end of the quadrant, thus corresponded to the fixed stars or zero force; intermediate ones were to be found proportional to the "forces" of the six planets. After a great amount of unfinished trial calculations, which took nearly a whole summer, he convinced himself that success did not lie that way. In July, 1595, while lecturing on the great planetary conjunctions, he drew quasi-triangles in a circular zodiac showing the slow progression of these points of conjunction at intervals of just over 240° or eight signs. The successive chords marked out a smaller circle to which they were tangents, about half the diameter of the zodiacal circle as drawn, and Kepler at once saw a similarity to the orbits of Saturn and Jupiter, the radius of the inscribed circle of an equilateral triangle being half that of the circumscribed circle. His natural sequence of ideas impelled him to try a square, in the hope that the circumscribed and inscribed circles might give him a similar "analogy" for the orbits of Jupiter and Mars. He next tried a pentagon and so on, but he soon noted that he would never reach the sun that way, nor would he find any such limitation as six, the number of "possibles" being obviously infinite. The actual planets moreover were not even six but only five, so far as he knew, so he next pondered the question of what sort of things these could be of which only five different figures were possible and suddenly thought of the five regular solids.[2] He immediately pounced upon this idea and ultimately evolved the following scheme. "The earth is the sphere, the measure of all; round it describe a dodecahedron; the sphere including this will be Mars. Round Mars describe a tetrahedron; the sphere including this will be Jupiter. Describe a cube round Jupiter; the sphere including this will be Saturn. Now, inscribe in the earth an icosahedron, the sphere inscribed in it will be Venus: inscribe an octahedron in Venus: the circle inscribed in it will be Mercury." With this result Kepler was inordinately pleased, and regretted not a moment of the time spent in obtaining it, though to us this "Mysterium Cosmographicum" can only appear useless, even without the more recent additions to the known planets. He admitted that a certain thickness must be assigned to the intervening spheres to cover the greatest and least distances of the several planets from the sun, but even then some of the numbers obtained are not a very close fit for the corresponding planetary orbits. Kepler's own suggested explanation of the discordances was that they must be due to erroneous measures of the planetary distances, and this, in those days of crude and infrequent observations, could not easily be disproved. He next thought of a variety of reasons why the five regular solids should occur in precisely the order given and in no other, diverging from this into a subtle and not very intelligible process of reasoning to account for the division of the zodiac into 360°. The next subject was more important, and dealt with the relation between the distances of the planets and their times of revolution round the sun. It was obvious that the period was not simply proportional to the distance, as the outer planets were all too slow for this, and he concluded "either that the moving intelligences of the planets are weakest in those that are farthest from the sun, or that there is one moving intelligence in the sun, the common centre, forcing them all round, but those most violently which are nearest, and that it languishes in some sort and grows weaker at the most distant, because of the remoteness and the attenuation of the virtue". This is not so near a guess at the theory of gravitation as might be supposed, for Kepler imagined that a repulsive force was necessary to account for the planets being sometimes further from the sun, and so laid aside the idea of a constant attractive force. He made several other attempts to find a law connecting the distances and periods of the planets, but without success at that time, and only desisted when by unconsciously arguing in a circle he appeared to get the same result from two totally different hypotheses. He sent copies of his book to several leading astronomers, of whom Galileo praised his ingenuity and good faith, while Tycho Brahe was evidently much struck with the work and advised him to adapt something similar to the Tychonic system instead of the Copernican. He also intimated that his Uraniborg observations would provide more accurate determinations of the planetary orbits, and thus made Kepler eager to visit him, a project which as we shall see was more than fulfilled. Another copy of the book Kepler sent to Reymers the Imperial astronomer with a most fulsome letter, which Tycho, who asserted that Reymers had simply plagiarised his work, very strongly resented, thus drawing from Kepler a long letter of apology. About the same time Kepler had married a lady already twice widowed, and become involved in difficulties with her relatives on financial grounds, and with the Styrian authorities in connection with the religious disputes then coming to a head. On account of these latter he thought it expedient, the year after his marriage, to withdraw to Hungary, from whence he sent short treatises to Tübingen, "On the magnet" (following the ideas of Gilbert of Colchester), "On the cause of the obliquity of the ecliptic" and "On the Divine wisdom as shown in the Creation". His next important step makes it desirable to devote a chapter to a short notice of Tycho Brahe.

> **Footnote 2:** Since the sum of the plane angles at a corner of a regular solid must be less than four right angles, it is easily seen that few regular solids are possible. Hexagonal faces are clearly impossible, or any polygonal faces with more than five sides. The possible forms are the dodecahedron with twelve pentagonal faces, three meeting at each corner; the cube, six square faces, three meeting at each corner; and three figures with triangular faces, the tetrahedron of four faces, three meeting at each corner; the octahedron of eight faces, four meeting at each corner; and the icosahedron of twenty faces, five meeting at each corner.

Chapter III
Tycho Brahe

The age following that of Copernicus produced three outstanding figures associated with the science of astronomy, then reaching the close of what Professor Forbes so aptly styles the geometrical period. These three Sir David Brewster has termed "Martyrs of Science"; Galileo, the great Italian philosopher, has his own place among the "Pioneers of Science"; and invaluable though Tycho Brahe's work was, the latter can hardly be claimed as a pioneer in the same sense as the other two. Nevertheless, Kepler, the third member of the trio, could not have made his most valuable discoveries without Tycho's observations.

Of noble family, born a twin on 14th December, 1546, at Knudstrup in Scania (the southernmost part of Sweden, then forming part of the kingdom of Denmark), Tycho was kidnapped a year later by a childless uncle. This uncle brought him up as his own son, provided him at the age of seven with a tutor, and sent him in 1559 to the University of Copenhagen, to study for a political career by taking courses in rhetoric and philosophy. On 21st August, 1560, however, a solar eclipse took place, total in Portugal, and therefore of small proportions in Denmark, and Tycho's keen interest was awakened, not so much by the phenomenon, as by the fact that it had occurred according to prediction. Soon afterwards he purchased an edition of Ptolemy in order to read up the subject of astronomy, to which, and to mathematics, he devoted most of the remainder of his three years' course at Copenhagen. His uncle next sent him to Leipzig to study law, but he managed to continue his astronomical researches. He obtained the Alphonsine and the new Prutenic Tables, but soon found that the latter, though more accurate than the former, failed to represent the true positions of the planets, and grasped the fact that continuous observation was essential in order to determine the true motions. He began by observing a conjunction of Jupiter and Saturn in August, 1563, and found the Prutenic Tables several days in error, and the Alphonsine a whole month. He provided himself with a cross-staff for determining the angular distance between stars or other objects, and, finding the divisions of the scale inaccurate, constructed a table of corrections, an improvement that seems to have been a decided innovation, the previous practice having been to use the best available instrument and ignore its errors. About this time war broke out between Denmark and Sweden, and Tycho returned to his uncle, who was vice-admiral and attached to the king's suite. The uncle died in the following month, and early in the next year Tycho went abroad again, this time to Wittenberg. After five months, however, an outbreak of plague drove him away, and he matriculated at Rostock, where he found little astronomy but a good deal of astrology. While there he fought a duel in the dark and lost part of his nose, which he replaced by a composition of gold and silver. He carried on regular observations with his cross-staff and persevered with his astronomical studies in spite of the objections and want of sympathy of his fellow-countrymen. The King of Denmark, however, having a higher opinion of the value of science, promised Tycho the first canonry that should fall vacant in the cathedral chapter of Roskilde, so that he might be assured of an income while devoting himself to financially unproductive work. In 1568 Tycho left Rostock, and matriculated at Basle, but soon moved on to Augsburg, where he found more enthusiasm for astronomy, and induced one of his new friends to order the construction of a large 19-foot quadrant of heavy oak beams. This was the first of the series of great instruments associated with Tycho's name, and it remained in use for five years, being destroyed by a great storm in 1574. Tycho meanwhile had left Augsburg in 1570 and returned to live with his father, now governor of Helsingborg Castle, until the latter's death in the following year. Tycho then joined his mother's brother, Steen Bille, the only one of his relatives who showed any sympathy with his desire for a scientific career.

On 11th November, 1572, Tycho noticed an unfamiliar bright star in the constellation of Cassiopeia, and continued to observe it with a sextant. It was a very brilliant object, equal to Venus at its brightest for the rest of November, not falling below the first magnitude for another four months, and remaining visible for more than a year afterwards. Tycho wrote a little book on the new star, maintaining that it had practically no parallax, and therefore could not be, as some supposed, a comet. Deeming authorship beneath the dignity of a noble he was very reluctant to publish, but he was convinced of the importance of increasing the number and accuracy of observations, though he was by no means free from all the erroneous ideas of his time. The little book contained a certain amount of astrology, but Tycho evidently did not regard this as of very great importance. He adopted the view that the very rarity of the phenomenon of a new star must prevent the formulation and adoption of definite rules for determining its significance. We gather from lectures which he was persuaded to deliver at the University of Copenhagen that, though in agreement with the accepted canons of astrology as to the influence of planetary conjunctions and such phenomena on the course of human events, he did not consider the fate predicted by anyone's horoscope to be unavoidable, but thought the great value of astrology lay in the warnings derived from such computations, which should enable the believer to avoid threatened calamities. In 1575 he left Denmark once more and made his way to Cassel, where he found a kindred spirit in the studious Landgrave, William IV. of Hesse, whose astronomical pursuits had been interrupted by his accession to the government of Hesse, in 1567. Tycho observed with him for some time, the two forming a firm friendship, and then visited successively Frankfort, Basle, and Venice, returning by way of Augsburg, Ratisbon, and Saalfeld to Wittenberg; on the way he acquired various astronomical manuscripts, made friends among practical astronomers, and examined new instruments. He seemed to have considered the advantages of the several places thus visited and decided on Basle, but on his return to Denmark to fetch his family with the object of transferring them to Basle, he found that his friend the Landgrave had written to King Frederick on his behalf, urging him to provide the means to enable Tycho to pursue his astronomical work, promising that not only should credit result for the king and for Denmark but that science itself would be greatly advanced. The ultimate result of this letter was that after refusing various offers, Tycho accepted from the king a grant of the small island of Hveen, in the Sound, with a guaranteed income, in addition to a large sum from the treasury for building an observatory on the island, far removed from the distractions of court life. Here Tycho built his celebrated observatory of Uraniborg and began observations in December, 1576, using the large instruments then found necessary in order to attain the accuracy of observation which within the next half-century was to be so greatly facilitated by the invention of the telescope. Here also he built several smaller observing rooms, so that his pupils should be able to observe independently. For more than twenty years he continued his observations at Uraniborg, surrounded by his family, and attracting numerous pupils. His constant aim was to accumulate a large store of observations of a high order of accuracy, and thus to provide data for the complete reform of astronomy. As we have seen, few of the Danish nobles had any sympathy with Tycho's pursuits, and most of them strongly resented the continual expense borne by the King's treasury. Tycho moreover was so absorbed in his scientific pursuits that he would not take the trouble to be a good landlord, nor to carry out all the duties laid upon him in return for certain of his grants of income. His buildings included a chemical laboratory, and he was in the habit of making up elixirs for various medical purposes; these were quite popular, particularly as he made no charge for them. He seems to have been something of a homœopathist, for he recommends sulphur to cure infectious diseases "brought on by the sulphurous vapours of the Aurora Borealis"!

King Frederick, in consideration of various grants to Tycho, relied upon his assistance in scientific matters, and especially in astrological calculations; such as the horoscope of the heir apparent, Prince Christian, born in 1577, which has been preserved among Tycho's writings. There is, however, no known copy in existence of any of the series of annual almanacs with predictions which he prepared for the King. In November, 1577, appeared a bright comet, which Tycho carefully observed with his sextant, proving that it had no perceptible parallax, and must therefore be further off than the moon. He thus definitely overthrew the common belief in the atmospheric origin of comets, which he had himself hitherto shared. With increasing accuracy he observed several other comets, notably one in 1585, when he had a full equipment of instruments and a large staff of assistants. The year 1588, which saw the death of his royal benefactor, saw also the publication of a volume of Tycho's great work "Introduction to the New Astronomy". The first volume, devoted to the new star of 1572, was not ready, because the reduction of the observations involved so much research to correct the star places for refraction, precession, etc.; it was not completed in fact until Tycho's death, but the second volume, dealing with the comet of 1577, was printed at Uraniborg and some copies were issued in 1588. Besides the comet observations it included an account of Tycho's system of the world. He would not accept the Copernican system, as he considered the earth too heavy and sluggish to move, and also that the authority of Scripture was against such an hypothesis. He therefore assumed that the other planets revolved about the sun, while the sun, moon, and stars revolved about the earth as a centre. Geometrically this is much the same as the Copernican system, but physically it involves the grotesque demand that the whole system of stars revolves round our insignificant little earth every twenty-four hours. Since his previous small book on the comet, Tycho had evidently considered more fully its possible astrological significance, for he foretold a religious war, giving the date of its commencement, and also the rising of a great Protestant champion. These predictions were apparently fulfilled almost to the letter by the great religious wars that broke out towards the end of the sixteenth century, and in the person of Gustavus Adolphus.

King Frederick's death did not at first affect Tycho's position, for the new king, Christian, was only eleven years old, and for some years the council of regents included two of his supporters. After their deaths, however, his emoluments began to be cut down on the plea of economy, and as he took very little trouble to carry out any other than scientific duties it was easy enough for his enemies to find fault. One after another source of income was cut off, but he persevered with his scientific work, including a catalogue of stars. He had obtained plenty of good observations of 777 stars, but thought his catalogue should contain 1000 stars, so he hastily observed as many more as he could up to the time of his leaving Hveen, though even then he had not completed his programme. About the time that King Christian reached the age of eighteen, Tycho began to look about for a new patron, and to consider the prospects offered by transferring himself with his instruments and activities to the patronage of the Emperor Rudolph II. In 1597, when even his pension from the Royal treasury was cut off, he hurriedly packed up his instruments and library, and after a few weeks' sojourn at Copenhagen, proceeded to Rostock, in Mecklenburg, whence he sent an appeal to King Christian. It is possible that had he done this before leaving Hveen it might have had more effect, but it can be readily seen from the tone of the king's unfavourable reply that his departure was regarded as an aggravation of previous shortcomings. Driven from Rostock by the plague, Tycho settled temporarily at Wandsbeck, in Holstein, but towards the end of 1598 set out to meet the Emperor at Prague. Once more plague intervened and he spent some time at Dresden, afterwards going to Wittenberg for the winter. He ultimately reached Prague in June, 1599. Rudolph granted him a salary of at least 3000 florins, promising also to settle on him the first hereditary estate that should lapse to the Crown. He offered, moreover, the choice between three castles outside Prague, of which Tycho chose Benatek. There he set about altering the buildings in readiness for his instruments, for which he sent to Uraniborg. Before they reached him, after many vexatious delays, he had given up waiting for the funds promised for his building expenses, and removed from Benatek to Prague. It was during this interval that after considerable negotiation, Kepler, who had been in correspondence with Tycho, consented to join him as an assistant. Another assistant, Longomontanus, who had been with Tycho at Uraniborg, was finding difficulty with the long series of Mars observations, and it was arranged that he should transfer his energies to the lunar observations, leaving those of Mars for Kepler. Before very much could be done with them, however, Tycho died at the end of October, 1601. He may have regretted the peaceful island of Hveen, considering the troubles in which Bohemia was rapidly becoming involved, but there is little doubt that had it not been for his self-imposed exile, his observations would not have come into Kepler's hands, and their great value might have been lost. In any case it was at Uraniborg that the mass of observations was produced upon which the fame of Tycho Brahe rests. His own discoveries, though in themselves the most important made in astronomy for many centuries, are far less valuable than those for which his observations furnished the material. He discovered the third and fourth inequalities of the moon in longitude, called respectively the variation and the annual equation, also the variability of the motion of the moon's nodes and the inclination of its orbit to the ecliptic. He obtained an improved value of the constant of precession, and did good service by rejecting the idea that it was variable, an idea which, under the name of trepidation, had for many centuries been accepted. He discovered the effect of refraction, though only approximately its amount, and determined improved values of many other astronomical constants, but singularly enough made no determination of the distance of the sun, adopting instead the ancient and erroneous value given by Hipparchus.

His magnificent Observatory of Uraniborg, the finest building for astronomical purposes that the world had hitherto seen, was allowed to fall into decay, and scarcely more than mere indications of the site may now be seen.

Chapter IV
Kepler Joins Tycho

The association of Kepler with Tycho was one of the most important landmarks in the history of astronomy. The younger man hoped, by the aid of Tycho's planetary observations, to obtain better support for some of his fanciful speculative theories, while the latter, who had certainly not gained in prestige by leaving Denmark, was in great need of a competent staff of assistants. Of the two it would almost seem that Tycho thought himself the greater gainer, for in spite of his reputation for brusqueness and want of consideration, he not only made light of Kepler's apology in the matter of Reymers, but treated him with uniform kindness in the face of great rudeness and ingratitude. He begged him to come "as a welcome friend," though Kepler, very touchy on the subject of his own astronomical powers, was afraid he might be regarded as simply a subordinate assistant. An arrangement had been suggested by which Kepler should obtain two years' leave of absence from Gratz on full pay, which, because of the higher cost of living in Prague, should be supplemented by the Emperor; but before this could be concluded, Kepler threw up his professorship, and thinking he had thereby also lost the chance of going to Prague, applied to Maestlin and others of his Tübingen friends to make interest for him with the Duke of Wurtemberg and secure the professorship of medicine. Tycho, however, still urged him to come to Prague, promising to do his utmost to secure for him a permanent appointment, or in any event to see that he was not the loser by coming. Kepler was delayed by illness on the way, but ultimately reached Prague, accompanied by his wife, and for some time lived entirely at Tycho's expense, writing by way of return essays against Reymers and another man, who had claimed the credit of the Tychonic system. This Kepler could do with a clear conscience, as it was only a question of priority and did not involve any support of the system, which he deemed far inferior to that of Copernicus. The following year saw friction between the two astronomers, and we learn from Kepler's abject letter of apology that he was entirely in the wrong. It was about money matters, which in one way or another embittered the rest of Kepler's life, and it arose during his absence from Prague. On his return in September, 1601, Tycho presented him to the Emperor, who gave him the title of Imperial Mathematician, on condition of assisting Tycho in his calculations, the very thing Kepler was most anxious to be allowed to do: for nowhere else in the world was there such a collection of good observations sufficient for his purpose of reforming the whole theory of astronomy. The Emperor's interest was still mainly with astrology, but he liked to think that his name would be handed down to posterity in connection with the new Planetary Tables in the same way as that of Alphonso of Castile, and he made liberal promises to pay the expenses. Tycho's other principal assistant, Longomontanus, did not stay long after giving up the Mars observations to Kepler, but instead of working at the new lunar theory, suddenly left to take up a professorship of astronomy in his native Denmark. Very shortly afterwards Tycho himself died of acute distemper; Kepler began to prepare the mass of manuscripts for publication, but, as everything was claimed by the Brahe family, he was not allowed to finish the work. He succeeded to Tycho's post of principal mathematician to the Emperor, at a reduced official salary, which owing to the emptiness of the Imperial treasury was almost always in arrear. In order to meet his expenses he had recourse to the casting of nativities, for which he gained considerable reputation and received very good pay. He worked by the conventional rules of astrology, and was quite prepared to take fees for so doing, although he had very little faith in them, preferring his own fanciful ideas.

In 1604 the constellation of Cassiopeia was once more temporarily enriched by the appearance of a new star, said by some to be brighter than Tycho's nova, and by others to be twice as bright as Jupiter. Kepler at once wrote a short account of it, from which may be gathered some idea of his attitude towards astrology. Contrasting the two novae, he says: "Yonder one chose for its appearance a time no way remarkable, and came into the world quite unexpectedly, like an enemy storming a town and breaking into the market-place before the citizens are aware of his approach; but ours has come exactly in the year of which astrologers have written so much about the fiery trigon that happens in it; just in the month in which (according to Cyprian), Mars comes up to a very perfect conjunction with the other two superior planets; just in the day when Mars has joined Jupiter, and just in the region where this conjunction has taken place. Therefore the apparition of this star is not like a secret hostile irruption, as was that one of 1572, but the spectacle of a public triumph, or the entry of a mighty potentate; when the couriers ride in some time before to prepare his lodgings, and the crowd of young urchins begin to think the time over long to wait, then roll in, one after another, the ammunition and money, and baggage waggons, and presently the trampling of horse and the rush of people from every side to the streets and windows; and when the crowd have gazed with their jaws all agape at the troops of knights; then at last the trumpeters and archers and lackeys so distinguish the person of the monarch, that there is no occasion to point him out, but every one cries of his own accord—'Here we have him'. What it may portend is hard to determine, and this much only is certain, that it comes to tell mankind either nothing at all or high and mighty news, quite beyond human sense and understanding. It will have an important influence on political and social relations; not indeed by its own nature, but as it were accidentally through the disposition of mankind. First, it portends to the booksellers great disturbances and tolerable gains; for almost every *Theologus, Philosophicus, Medicus,* and *Mathematicus,* or whoever else, having no laborious occupation entrusted to him, seeks his pleasure *in studiis,* will make particular remarks upon it, and will wish to bring these remarks to the light. Just so will others, learned and unlearned, wish to know its meaning, and they will buy the authors who profess to tell them. I mention these things merely by way of example, because although thus much can be easily predicted without great skill, yet may it happen just as easily, and in the same manner, that the vulgar, or whoever else is of easy faith, or, it may be, crazy, may wish to exalt himself into a great prophet; or it may even happen that some powerful lord, who has good foundation and beginning of great dignities, will be cheered on by this phenomenon to venture on some new scheme, just as if God had set up this star in the darkness merely to enlighten them." He made no secret of his views on conventional astrology, as to which he claimed to speak with the authority of one fully conversant with its principles, but he nevertheless expressed his sincere conviction that the conjunctions and aspects of the planets certainly did affect things on the earth, maintaining that he was driven to this belief against his will by "most unfailing experiences".

Meanwhile the projected Rudolphine Tables were continually delayed by the want of money. Kepler's nominal salary should have been ample for his expenses, increased though they were by his growing family, but in the depleted state of the treasury there were many who objected to any payment for such "unpractical" purposes. This particular attitude has not been confined to any special epoch or country, but the obvious result in Kepler's case was to compel him to apply himself to less expensive matters than the Planetary Tables, and among these must be included not only the horoscopes or nativities, which owing to his reputation were always in demand, but also other writings which probably did not pay so well. In 1604 he published "A Supplement to Vitellion," containing the earliest known reasonable theory of optics, and especially of dioptrics or vision through lenses. He compared the mechanism of the eye with that of Porta's "Camera

Obscura," but made no attempt to explain how the image formed on the retina is understood by the brain. He went carefully into the question of refraction, the importance of which Tycho had been the first astronomer to recognise, though he only applied it at low altitudes, and had not arrived at a true theory or accurate values. Kepler wasted a good deal of time and ingenuity on trial theories. He would invariably start with some hypothesis, and work out the effect. He would then test it by experiment, and when it failed would at once recognise that his hypothesis was *a priori* bound to fail. He rarely seems to have noticed the fatal objections in time to save himself trouble. He would then at once start again on a new hypothesis, equally gratuitous and equally unfounded. It never seems to have occurred to him that there might be a better way of approaching a problem. Among the lines he followed in this particular investigation were, first, that refraction depends only on the angle of incidence, which, he says, cannot be correct as it would thus be the same for all refracting substances; next, that it depended also on the density of the medium. This was a good shot, but he unfortunately assumed that all rays passing into a denser medium would apparently penetrate it to a depth depending only on the medium, which means that there is a constant ratio between the tangents, instead of the sines, of the inclination of the incident and refracted rays to the normal. Experiment proved that this gave too high values for refraction near the vertical compared with those near the horizon, so Kepler "went off at a tangent" and tried a totally new set of ideas, which all reduced to the absurdity of a refraction which vanished at the horizon. These were followed by another set, involving either a constant amount of refraction or one becoming infinite. He then came to the conclusion that these geometrical methods must fail because the refracted image is not real, and determined to try by analogy only, comparing the equally unreal image formed by a mirror with that formed by refraction in water. He noticed how the bottom of a vessel containing water appears to rise more and more away from the vertical, and at once jumped to the analogy of a concave mirror, which magnifies the image, while a convex mirror was likened to a rarer medium. This line of attack also failed him, as did various attempts to find relations between his measurements of refraction and conic sections, and he broke off suddenly with a diatribe against Tycho's critics, whom he likened to blind men disputing about colours. Not many years later Snell discovered the true law of refraction, but Kepler's contribution to the subject, though he failed to discover the actual law, includes several of the adopted "by-laws". He noted that atmospheric refraction would alter with the height of the atmosphere and with temperature, and also recognised the fact that rainbow colours depend on the angle of refraction, whether seen in the rainbow itself, or in dew, glass, water, or any similar medium. He thus came near to anticipating Newton. Before leaving the subject of

Kepler's optics it will be well to recall that a few years later after hearing of Galileo's telescope, Kepler suggested that for astronomical purposes two convex lenses should be used, so that there should be a real image where measuring wires could be placed for reference. He did not carry out the idea himself, and it was left to the Englishman Gascoigne to produce the first instrument on this "Keplerian" principle, universally known as the Astronomical Telescope.

In 1606 came a second treatise on the new star, discussing various theories to account for its appearance, and refusing to accept the notion that it was a "fortuitous concourse of atoms". This was followed in 1607 by a treatise on comets, suggested by the comet appearing that year, known as Halley's comet after its next return. He regarded comets as "planets" moving in straight lines, never having examined sufficient observations of any comet to convince himself that their paths are curved. If he had not assumed that they were external to the system and so could not be expected to return, he might have anticipated Halley's discovery. Another suggestive remark of his was to the effect that the planets must be self-luminous, as otherwise Mercury and Venus, at any rate, ought to show phases. This was put to the test not long afterwards by means of Galileo's telescope.

In 1607 Kepler rushed into print with an alleged observation of Mercury crossing the sun, but after Galileo's discovery of sun-spots, Kepler at once cheerfully retracted his observation of "Mercury," and so far was he from being annoyed or bigoted in his views, that he warmly adopted Galileo's side, in contrast to most of those whose opinions were liable to be overthrown by the new discoveries. Maestlin and others of Kepler's friends took the opposite view.

Chapter V
Kepler's Laws

When Gilbert of Colchester, in his "New Philosophy," founded on his researches in magnetism, was dealing with tides, he did not suggest that the moon attracted the water, but that "subterranean spirits and humours, rising in sympathy with the moon, cause the sea also to rise and flow to the shores and up rivers". It appears that an idea, presented in some such way as this, was more readily received than a plain statement. This so-called philosophical method was, in fact, very generally applied, and Kepler, who shared Galileo's admiration for Gilbert's work, adopted it in his own attempt to extend the idea of magnetic attraction to the planets. The general idea of "gravity" opposed the hypothesis of the rotation of the earth on the ground that loose objects would fly off: moreover, the latest refinements of the old system of planetary motions necessitated their orbits being described about a mere empty point. Kepler very strongly combated these notions, pointing out the absurdity of the conclusions to which they tended, and proceeded in set terms to describe his own theory.

"Every corporeal substance, so far forth as it is corporeal, has a natural fitness for resting in every place where it may be situated by itself beyond the sphere of influence of a body cognate with it. Gravity is a mutual affection between cognate bodies towards union or conjunction (similar in kind to the magnetic virtue), so that the earth attracts a stone much rather than the stone seeks the earth. Heavy bodies (if we begin by assuming the earth to be in the centre of the world) are not carried to the centre of the world in its quality of centre of the world, but as to the centre of a cognate round body, namely, the earth; so that wheresoever the earth may be placed, or whithersoever it may be carried by its animal faculty, heavy bodies will always be carried towards it. If the earth were not round, heavy bodies would not tend from every side in a straight line towards the centre of the earth, but to different points from different sides. If two stones were placed in any part of the world near each other, and beyond the sphere of influence of a third cognate body, these stones, like two magnetic needles, would come together in the intermediate point, each approaching the other by a space proportional to the comparative mass of the other. If the moon and earth were not retained in their orbits by their animal force or some other equivalent, the earth would mount to the moon by a fifty-fourth part of their distance, and the moon fall towards the earth through the other fifty-three parts, and they would there meet, assuming, however, that the substance of both is of the same density. If the earth should cease to attract its waters to itself all the waters of the sea would he raised and would flow to the body of the moon. The sphere of the attractive virtue which is in the moon extends as far as the earth, and entices up the waters; but as the moon flies rapidly across the zenith, and the waters cannot follow so quickly, a flow of the ocean is occasioned in the torrid zone towards the westward. If the attractive virtue of the moon extends as far as the earth, it follows with greater reason that the attractive virtue of the earth extends as far as the moon and much farther; and, in short, nothing which consists of earthly substance anyhow constituted although thrown up to any height, can ever escape the powerful operation of this attractive virtue. Nothing which consists of corporeal matter is absolutely light, but that is comparatively lighter which is rarer, either by its own nature, or by accidental heat. And it is not to be thought that light bodies are escaping to the surface of the universe while they are carried upwards, or that they are not attracted by the earth. They are attracted, but in a less degree, and so are driven outwards by the heavy bodies; which being done, they stop, and are kept by the earth in their own place. But although the attractive virtue of the earth extends upwards, as has been said, so very far, yet if any stone should be at a distance great enough to become sensible compared with the earth's diameter, it is true that on the motion of the earth such a stone would not follow altogether; its own force of resistance would be combined with the attractive force of the earth, and thus it would extricate itself in some degree from the motion of the earth." The above passage from the Introduction to Kepler's "Commentaries on the Motion of Mars," always regarded as his most valuable work, must have been known to Newton, so that no such incident as the fall of an apple was required to provide a necessary and sufficient explanation of the genesis of his Theory of Universal Gravitation. Kepler's glimpse at such a theory could have been no more than a glimpse, for he went no further with it. This seems a pity, as it is far less fanciful than many of his ideas, though not free from the "virtues" and "animal faculties," that correspond to Gilbert's "spirits and humours". We must, however, proceed to the subject of Mars, which was, as before noted, the first important investigation entrusted to Kepler on his arrival at Prague.

The time taken from one opposition of Mars to the next is decidedly unequal at different parts of his orbit, so that many oppositions must be used to determine the mean motion. The ancients had noticed that what was called the "second inequality," due as we now know to the orbital motion of the earth, only vanished when earth, sun, and planet were in line, i.e. at the planet's opposition; therefore they used oppositions to determine the mean motion, but deemed it necessary to apply a correction to the true opposition to reduce to mean opposition, thus sacrificing part of the advantage of using oppositions. Tycho and Longomontanus had followed this method in their calculations from Tycho's twenty years' observations. Their aim was to find a position of the "equant," such that these observations would show a constant angular motion about it; and that the computed positions would agree in latitude and longitude with the actual observed positions. When Kepler arrived he was told that their longitudes agreed within a couple of minutes of arc, but that something was wrong with the latitudes. He found, however, that even in longitude their positions showed discordances ten times as great as they admitted, and so, to clear the ground of assumptions as far as possible, he determined to use true oppositions. To this Tycho objected, and Kepler had great difficulty in convincing him that the new move would be any improvement, but undertook to prove to him by actual examples that a false position of the orbit could by adjusting the equant be made to fit the longitudes within five minutes of arc, while giving quite erroneous values of the latitudes and second inequalities. To avoid the possibility of further objection he carried out this demonstration separately for each of the systems of Ptolemy, Copernicus, and Tycho. For the new method he noticed that great accuracy was required in the reduction of the observed places of Mars to the ecliptic, and for this purpose the value obtained for the parallax by Tycho's assistants fell far short of the requisite accuracy. Kepler therefore was obliged to recompute the parallax from the original observations, as also the position of the line of nodes and the inclination of the orbit. The last he found to be constant, thus corroborating his theory that the plane of the orbit passed through the sun. He repeated his calculations no fewer than seventy times (and that before the invention of logarithms), and at length adopted values for the mean longitude and longitude of aphelion. He found no discordance greater than two minutes of arc in Tycho's observed longitudes in opposition, but the latitudes, and also longitudes in other parts of the orbit were much more discordant, and he found to his chagrin that four years' work was practically wasted. Before making a fresh start he looked for some simplification of the labour; and determined to adopt Ptolemy's assumption known as the principle of the bisection of the excentricity. Hitherto, since Ptolemy had given no reason for this assumption, Kepler had preferred not to make it, only taking for granted that the centre was at some point on the line called the excentricity (see Figs. 1, 2).

A marked improvement in residuals was the result of this step, proving, so far, the correctness of Ptolemy's principle, but there still remained discordances amounting to eight minutes of arc. Copernicus, who had no idea of the accuracy obtainable in observations, would probably have regarded such an agreement as remarkably good; but Kepler refused to admit the possibility of an error of eight minutes in any of Tycho's observations. He thereupon vowed to construct from these eight minutes a new planetary theory that should account for them all. His repeated failures had by this time convinced him that no uniformly described circle could possibly represent the motion of Mars. Either the orbit could not be circular, or else the angular velocity could not be constant about any point whatever. He determined to attack the "second inequality," i.e. the optical illusion caused by the earth's annual motion, but first revived an old idea of his own that for the sake of uniformity the sun, or as he preferred to regard it, the earth, should have an equant as well as the planets. From the irregularities of the solar motion he soon found that this was the case, and that the motion was uniform about a point on the line from the sun to the centre of the earth's orbit, such that the centre bisected the distance from the sun to the "Equant"; this fully supported Ptolemy's principle. Clearly then the earth's linear velocity could not be constant, and Kepler was encouraged to revive another of his speculations as to a force which was weaker at greater distances. He found the velocity greater at the nearer apse, so that the time over an equal arc at either apse was proportional to the distance. He conjectured that this might prove to be true for arcs at all parts of the orbit, and to test this he divided the orbit into 360 equal parts, and calculated the distances to the points of division. Archimedes had obtained an approximation to the area of a circle by dividing it radially into a very large number of triangles, and Kepler had this device in mind. He found that the sums of successive distances from his 360 points were approximately proportional to the times from point to point, and was thus enabled to represent much more accurately the annual motion of the earth which produced the second inequality of Mars, to whose motion he now returned. Three points are sufficient to define a circle, so he took three observed positions of Mars and found a circle; he then took three other positions, but obtained a different circle, and a third set gave yet another. It thus began to appear that the orbit could not be a circle. He next tried to divide into 360 equal parts, as he had in the case of the earth, but the sums of distances failed to fit the times, and he realised that the sums of distances were not a good measure of the area of successive triangles. He noted, however, that the errors at the apses were now smaller than with a central circular orbit, and of the opposite sign, so he determined to try whether an oval orbit would fit better, following a suggestion made by Purbach in the case of Mercury, whose orbit is even more eccentric than that of Mars, though observations were too scanty to form the foundation of any theory. Kepler gave his fancy play in the choice of an oval, greater at one end than the other, endeavouring to satisfy some ideas about epicyclic motion, but could not find a satisfactory curve. He then had the fortunate idea of trying an ellipse with the same axis as his tentative oval. Mars now appeared too slow at the apses instead of too quick, so obviously some intermediate ellipse must be sought between the trial ellipse and the circle on the same axis. At this point the "long arm of coincidence" came into play. Half-way between the apses lay the mean distance, and at this position the error was half the distance between the ellipse and the circle, amounting to .00429 of a radius. With these figures in his mind, Kepler looked up the greatest optical inequality of Mars, the angle between the straight lines from Mars to the Sun and to the centre of the circle.[3] The secant of this angle was 1.00429, so that he noted that an ellipse reduced from the circle in the ratio of 1.00429 to 1 would fit the motion of Mars at the mean distance as well as the apses.

> **Footnote 3:** This is clearly a maximum at AMC in Fig. 2,
> when its tangent AC / CM = the eccentricity.

It is often said that a coincidence like this only happens to somebody who "deserves his luck," but this simply means that recognition is essential to the coincidence. In the same way the appearance of one of a large number of people mentioned is hailed as a case of the old adage "Talk of the devil, etc.," ignoring all the people who failed to appear. No one, however, will consider Kepler unduly favoured. His genius, in his case certainly "an infinite capacity for taking pains," enabled him out of his medley of hypotheses, mainly unsound, by dint of enormous labour and patience, to arrive thus at the first two of the laws which established his title of "Legislator of the Heavens".

Figures Explanatory of Kepler's Theory of the Motion of Mars.

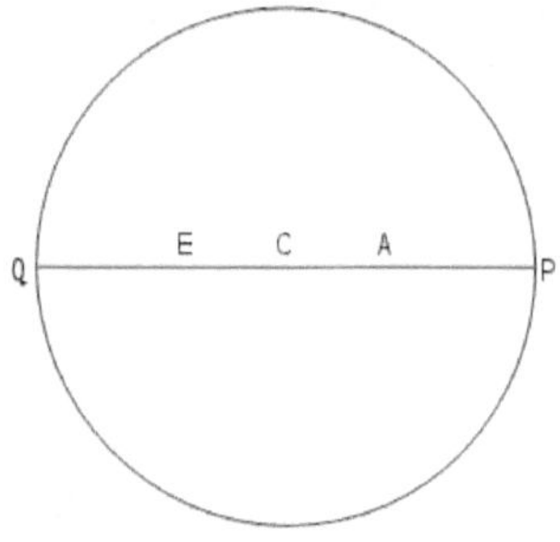

Fig. 1.

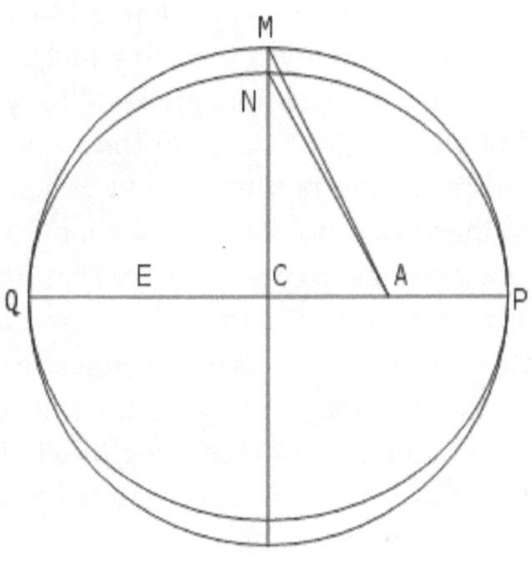

Fig. 2.

Fig. 1.—In Ptolemy's excentric theory, A may be taken to represent the earth, C the centre of a planet's orbit, and E the equant, P (perigee) and Q (apogee) being the apses of the orbit. Ptolemy's idea was that uniform motion in a circle must be provided, and since the motion was not uniform about the earth, A could not coincide with C; and since the motion still failed to be uniform about A or C, some point E must be found about which the motion should be uniform.

Fig. 2.—This is not drawn to scale, but is intended to illustrate Kepler's modification of Ptolemy's excentric. Kepler found velocities at P and Q proportional not to AP and AQ but to AQ and AP, or to EP and EQ if EC = CA (bisection of the excentricity). The velocity at M was wrong, and AM appeared too great. Kepler's first ellipse had M moved too near C. The distance AC is much exaggerated in the figure, as also is MN. AN = CP, the radius of the circle. MN should be .00429 of the radius, and MC / NC should be 1.00429. The velocity at N appeared to be proportional to EN (= AN). Kepler concluded that Mars moved round PNQ, so that the area described about A (the sun) was equal in equal times, A being the focus of the ellipse PNQ. The angular velocity is not quite constant about E, the equant or empty focus, but the difference could hardly have been detected in Kepler's time.

Kepler's improved determination of the earth's orbit was obtained by plotting the different positions of the earth corresponding to successive rotations of Mars, i.e. intervals of 687 days. At each of these the date of the year would give the angle MSE (Mars-Sun-Earth), and Tycho's observation the angle MES. So the triangle could be solved except for scale, and the ratio of SE to SM would give the distance of Mars from the sun in terms of that of the earth. Measuring from a fixed position of Mars (e.g. perihelion), this gave the variation of SE, showing the earth's inequality. Measuring from a fixed position of the earth, it would give similarly a series of positions of Mars, which, though lying not far from the circle whose diameter was the axis of Mars' orbit, joining perihelion and aphelion, always fell inside the circle except at those two points. It was a long time before it dawned upon Kepler that the simplest figure falling within the circle except at the two extremities of the diameter, was an ellipse, and it is not clear why his first attempt with an ellipse should have been just as much too narrow as the circle was too wide. The fact remains that he recognised suddenly that halving this error was tantamount to reducing the circle to the ellipse whose eccentricity was that of the old theory, i.e. that in which the sun would be in one focus and the equant in the other.

Having now fitted the ends of both major and minor axes of the ellipse, he leaped to the conclusion that the orbit would fit everywhere.

The practical effect of his clearing of the "second inequality" was to refer the orbit of Mars directly to the sun, and he found that the area between successive distances of Mars from the sun (instead of the sum of the distances) was strictly proportional to the time taken, in short, equal areas were described in equal times (2nd Law) when referred to the sun in the focus of the ellipse (1st Law).

He announced that (1) The planet describes an ellipse, the sun being in one focus; and (2) The straight line joining the planet to the sun sweeps out equal areas in any two equal intervals of time. These are Kepler's first and second Laws though not discovered in that order, and it was at once clear that Ptolemy's "bisection of the excentricity" simply amounted to the fact that the centre of an ellipse bisects the distance between the foci, the sun being in one focus and the angular velocity being uniform about the empty focus. For so many centuries had the fetish of circular motion postponed discovery. It was natural that Kepler should assume that his laws would apply equally to all the planets, but the proof of this, as well as the reason underlying the laws, was only given by Newton, who approached the subject from a totally different standpoint.

This commentary on Mars was published in 1609, the year of the invention of the telescope, and Kepler petitioned the Emperor for further funds to enable him to complete the study of the other planets, but once more there was delay; in 1612 Rudolph died, and his brother Matthias who succeeded him, cared very little for astronomy or even astrology, though Kepler was reappointed to his post of Imperial Mathematician. He left Prague to take up a permanent professorship at the University of Linz.

His own account of the circumstances is gloomy enough. He says, "In the first place I could get no money from the Court, and my wife, who had for a long time been suffering from low spirits and despondency, was taken violently ill towards the end of 1610, with the Hungarian fever, epilepsy and phrenitis. She was scarcely convalescent when all my three children were at once attacked with smallpox. Leopold with his army occupied the town beyond the river just as I lost the dearest of my sons, him whose nativity you will find in my book on the new star. The town on this side of the river where I lived was harassed by the Bohemian troops, whose new levies were insubordinate and insolent; to complete the whole, the Austrian army brought the plague with them into the city. I went into Austria and endeavoured to procure the situation which I now hold. Returning in June, I found my wife in a decline from her grief at the death of her son, and on the eve of an infectious fever, and I lost her also within eleven days of my return. Then came fresh annoyance, of course, and her fortune was to be divided with my step-sisters. The Emperor Rudolph would not agree to my departure; vain hopes were given me of being paid from Saxony; my time and money were wasted together, till on the death of the Emperor in 1612, I was named again by his successor, and suffered to depart to Linz."

Being thus left a widower with a ten-year-old daughter Susanna, and a boy Louis of half her age, he looked for a second wife to take charge of them. He has given an account of eleven ladies whose suitability he considered. The first, an intimate friend of his first wife, ultimately declined; one was too old, another an invalid, another too proud of her birth and quarterings, another could do nothing useful, and so on. Number eight kept him guessing for three months, until he tired of her constant indecision, and confided his disappointment to number nine, who was not impressed. Number ten, introduced by a friend, Kepler found exceedingly ugly and enormously fat, and number eleven apparently too young. Kepler then reconsidered one of the earlier ones, disregarding the advice of his friends who objected to her lowly station. She was the orphan daughter of a cabinetmaker, educated for twelve years by favour of the Lady of Stahrenburg, and Kepler writes of her: "Her person and manners are suitable to mine; no pride, no extravagance; she can bear to work; she has a tolerable knowledge of how to manage a family; middle-aged and of a disposition and capability to acquire what she still wants".

Wine from the Austrian vineyards was plentiful and cheap at the time of the marriage, and Kepler bought a few casks for his household. When the seller came to ascertain the quantity, Kepler noticed that no proper allowance was made for the bulging parts, and the upshot of his objections was that he wrote a book on a new method of gauging—one of the earliest specimens of modern analysis, extending the properties of plane figures to segments of cones and cylinders as being "incorporated circles". He was summoned before the Diet at Ratisbon to give his opinion on the Gregorian Reform of the Calendar, and soon afterwards was excommunicated, having fallen foul of the Roman Catholic party at Linz just as he had previously at Gratz, the reason apparently being that he desired to think for himself. Meanwhile his salary was not paid any more regularly than before, and he was forced to supplement it by publishing what he called a "vile prophesying almanac which is scarcely more respectable than begging unless it be because it saves the Emperor's credit, who abandons me entirely, and with all his frequent and recent orders in council, would suffer me to perish with hunger".

In 1617 he was invited to Italy to succeed Magini as Professor of Mathematics at Bologna. Galileo urged him to accept the post, but he excused himself on the ground that he was a German and brought up among Germans with such liberty of speech as he thought might get him into trouble in Italy. In 1619 Matthias died and was succeeded by Ferdinand III, who again retained Kepler in his post. In the same year Kepler reprinted his "Mysterium Cosmographicum," and also published his "Harmonics" in five books dedicated to James I of England. "The first geometrical, on the origin and demonstration of the laws of the figures which produce harmonious proportions; the second, architectonical, on figurate geometry and the congruence of plane and solid regular figures; the third, properly Harmonic, on the derivation of musical proportions from figures, and on the nature and distinction of things relating to song, in opposition to the old theories; the fourth, metaphysical, psychological, and astrological, on the mental essence of Harmonics, and of their kinds in the world, especially on the harmony of rays emanating on the earth from the heavenly bodies, and on their effect in nature and on the sublunary and human soul; the fifth, astronomical and metaphysical, on the very exquisite Harmonics of the celestial motions and the origin of the excentricities in harmonious proportions." The extravagance of his fancies does not appear until the fourth book, in which he reiterates the statement that he was forced to adopt his astrological opinions from direct and positive observation. He despises "The common herd of prophesiers who describe the operations of the stars as if they were a sort of deities, the lords of heaven and earth, and producing everything at their pleasure. They never trouble themselves to consider what means the stars have of working any effects among us on the earth whilst they remain in the sky and send down nothing to us which is obvious to the senses, except rays of light." His own notion is "Like one who listens to a sweet melodious song, and by the gladness of his countenance, by his voice, and by the beating of his hand or foot attuned to the music, gives token that he perceives and approves the harmony: just so does sublunary nature, with the notable and evident emotion of the bowels of the earth, bear like witness to the same feelings, especially at those times when the rays of the planets form harmonious configurations on the earth," and again "The earth is not an animal like a dog, ready at every nod; but more like a bull or an elephant, slow to become angry, and so much the more furious when incensed." He seems to have believed the earth to be actually a living animal, as witness the following: "If anyone who has climbed the peaks of the highest mountains, throw a stone down their very deep clefts, a sound is heard from them; or if he throw it into one of the mountain lakes, which beyond doubt are bottomless, a storm will immediately arise, just as when you thrust a straw into the ear or nose of a ticklish animal, it shakes its head, or runs shudderingly away. What so like breathing, especially of those fish who draw water into their mouths and spout it out again through their gills, as that wonderful tide! For although it is so regulated according to the course of the moon, that, in the preface to my 'Commentaries on Mars,' I have mentioned it as probable that the waters are attracted by the moon, as iron by the loadstone, yet if anyone uphold that the earth regulates its breathing according to the motion of the sun and moon, as animals have daily and nightly alternations of sleep and waking, I shall not think his philosophy unworthy of being listened to; especially if any flexible parts should be discovered in the depths of the earth, to supply the functions of lungs or gills."

In the same book Kepler enlarges again on his views in reference to the basis of astrology as concerned with nativities and the importance of planetary conjunctions. He gives particulars of his own nativity. "Jupiter nearest the nonagesimal had passed by four degrees the trine of Saturn; the Sun and Venus in conjunction were moving from the latter towards the former, nearly in sextiles with both: they were also removing from quadratures with Mars, to which Mercury was closely approaching: the moon drew near to the trine of the same planet, close to the Bull's Eye even in latitude. The 25th degree of Gemini was rising, and the 22nd of Aquarius culminating. That there was this triple configuration on that day—namely the sextile of Saturn and the Sun, the sextile of Mars and Jupiter, and the quadrature of Mercury and Mars, is proved by the change of weather; for after a frost of some days, that very day became warmer, there was a thaw and a fall of rain." This alleged "proof" is interesting as it relies on the same principle which was held to justify the correction of an uncertain birth-time, by reference to illnesses, etc., met with later. Kepler however goes on to say, "If I am to speak of the results of my studies, what, I pray, can I find in the sky, even remotely alluding to it? The learned confess that several not despicable branches of philosophy have been newly extricated or amended or brought to perfection by me: but here my constellations were, not Mercury from the East in the angle of the seventh, and in quadratures with Mars, but Copernicus, but Tycho Brahe, without whose books of observations everything now set by me in the clearest light must have remained buried in darkness; not Saturn predominating Mercury, but my lords the Emperors Rudolph and Matthias, not Capricorn the house of Saturn but Upper Austria, the house of the Emperor, and the ready and unexampled bounty of his nobles to my petition. Here is that corner, not the western one of the horoscope, but on the earth whither, by permission of my Imperial master, I have betaken myself from a too uneasy Court; and whence, during these years of my life, which now tends towards its setting, emanate these Harmonics and the other matters on which I am engaged."

The fifth book contains a great deal of nonsense about the harmony of the spheres; the notes contributed by the several planets are gravely set down, that of Mercury having the greatest resemblance to a melody, though perhaps more reminiscent of a bugle-call. Yet the book is not all worthless for it includes Kepler's Third Law, which he had diligently sought for years. In his own words, "The proportion existing between the periodic times of any two planets is exactly the sesquiplicate proportion of the mean distances of the orbits," or as generally given, "the squares of the periodic times are proportional to the cubes of the mean distances." Kepler was evidently transported with delight and wrote, "What I prophesied two and twenty years ago, as soon as I discovered the five solids among the heavenly orbits,—what I firmly believed long before I had seen Ptolemy's 'Harmonics'—what I had promised my friends in the title of this book, which I named before I was sure of my discovery,—what sixteen years ago I urged as a thing to be sought,—that for which I joined Tycho Brahe, for which I settled in Prague, for which I have devoted the best part of my life to astronomical computations, at length I have brought to light, and have recognised its truth beyond my most sanguine expectations. Great as is the absolute nature of Harmonics, with all its details as set forth in my third book, it is all found among the celestial motions, not indeed in the manner which I imagined (that is not the least part of my delight), but in another very different, and yet most perfect and excellent. It is now eighteen months since I got the first glimpse of light, three months since the dawn, very few days since the unveiled sun, most admirable to gaze on, burst out upon me. Nothing holds me; I will indulge in my sacred fury; I will triumph over mankind by the honest confession that I have stolen the golden vases of the Egyptians to build up a tabernacle for my God far away from the confines of Egypt. If you forgive me, I rejoice, if you are angry, I can bear it; the die is cast, the book is written; to be read either now or by posterity, I care not which; it may well wait a century for a reader, as God has waited six thousand years for an observer." He gives the date 15th May, 1618, for the completion of his discovery. In his "Epitome of the Copernican Astronomy," he gives his own idea as to the reason for this Third Law. "Four causes concur for lengthening the periodic time. First, the length of the path; secondly, the weight or quantity of matter to be carried; thirdly, the degree of strength of the moving virtue; fourthly, the bulk or space into which is spread out the matter to be moved. The orbital paths of the planets are in the simple ratio of the distances; the weights or quantities of matter in different planets are in the subduplicate ratio of the same distances, as has been already proved; so that with every increase of distance a planet has more matter and therefore is moved more slowly, and accumulates more time in its revolution, requiring already, as it did, more time by reason of the length of the way. The third and fourth causes compensate each other in a comparison of different planets; the simple and subduplicate proportion compound the sesquiplicate proportion, which therefore is the ratio of the periodic times." The only part of this "explanation" that is true is that the paths are in the simple ratio of the distances, the "proof" so confidently claimed being of the circular kind commonly known as "begging the question". It was reserved for Newton to establish the Laws of Motion, to find the law of force that would constrain a planet to obey Kepler's first and second Laws, and to prove that it must therefore also obey the third.

Chapter VI
Closing Years

Soon after its publication Kepler's "Epitome" was placed along with the book of Copernicus, on the list of books prohibited by the Congregation of the Index at Rome, and he feared that this might prevent the publication or sale of his books in Austria also, but was told that though Galileo's violence was getting him into trouble, there would be no difficulty in obtaining permission for learned men to read any prohibited books, and that he (Kepler) need fear nothing so long as he remained quiet.

In his various works on Comets, he adhered to the opinion that they travelled in straight lines with varying velocity. He suggested that comets come from the remotest parts of ether, as whales and monsters from the depth of the sea, and that perhaps they are something of the nature of silkworms, and are wasted and consumed in spinning their own tails. Napier's invention of logarithms at once attracted Kepler's attention. He must have regretted that the discovery was not made early enough to save him a vast amount of labour in computations, but he managed to find time to compute some logarithm tables for himself, though he does not seem to have understood quite what Napier had done, and though with his usual honesty he gave full credit to the Scottish baron for his invention.

Though Eugenists may find a difficulty in reconciling Napier's brilliancy with the extreme youth of his parents, they may at any rate attribute Kepler's occasional fits of bad temper to heredity. His cantankerous mother, Catherine Kepler, had for some years been carrying on an action for slander against a woman who had accused her of administering a poisonous potion. Dame Kepler employed a young advocate who for reasons of his own "nursed" the case so long that after five years had elapsed without any conclusion being reached another judge was appointed, who had himself suffered from the caustic tongue of the prosecutrix, and so was already prejudiced against her. The defendant, knowing this, turned the tables on her opponent by bringing an accusation of witchcraft against her, and Catherine Kepler was imprisoned and condemned to the torture in July, 1620. Kepler, hearing of the sentence, hurried back from Linz, and succeeded in stopping the completion of the sentence, securing his mother's release the following year, as it was made clear that the only support for the case against her was her own intemperate language. Kepler returned to Linz, and his mother at once brought another action for costs and damages against her late opponent, but died before the case could be tried.

A few months before this Sir Henry Wotton, English Ambassador to Venice, visited Kepler, and finding him as usual, almost penniless, urged him to go to England, promising him a warm welcome there. Kepler, however, would not at that time leave Germany, giving several reasons, one of which was that he dreaded the confinement of an island. Later on he expressed his willingness to go as soon as his Rudolphine Tables were published, and lecture on them, even in England, if he could not do it in Germany, and if a good enough salary were forthcoming.

In 1624 he went to Vienna, and managed to extract from the Treasury 6000 florins on account of expenses connected with the Tables, but, instead of a further grant, was given letters to the States of Swabia, which owed money to the Imperial treasury. Some of this he succeeded in collecting, but the Tables were still further delayed by the religious disturbances then becoming violent. The Jesuits contrived to have Kepler's library sealed up, and, but for the Imperial protection, would have imprisoned him also; moreover the peasants revolted and blockaded Linz. In 1627, however, the long promised Tables, the first to discard the conventional circular motion, were at last published at Ulm in four parts. Two of these parts consisted of subsidiary Tables, of logarithms and other computing devices, another contained Tables of the elements of the sun, moon, and planets, and the fourth gave the places of a thousand stars as determined by Tycho, with Tycho's refraction Tables, which had the peculiarity of using different values for the refraction of the sun, moon, and stars. From a map prefixed to some copies of the Tables, we may infer that Kepler was one of the first, if not actually the first, to suggest the method of determining differences of longitude by occultations of stars at the moon's limb. In an Appendix, he showed how his Tables could be used by astrologers for their predictions, saying "Astronomy is the daughter of Astrology, and this modern Astrology again is the daughter of Astronomy, bearing something of the lineaments of her grandmother; and, as I have already said, this foolish daughter, Astrology, supports her wise but needy mother, Astronomy, from the profits of a profession not generally considered creditable". There is no doubt that Kepler strongly resented having to depend so much for his income on such methods which he certainly did not consider creditable.

It was probably Galileo whose praise of the new Tables induced the Grand Duke of Tuscany to send Kepler a gold chain soon after their publication, and we may perhaps regard it as a mark of favour from the

Emperor Ferdinand that he permitted Kepler to attach himself to the great Wallenstein, now Duke of Friedland, and a firm believer in Astrology. The Duke was a better paymaster than either of the three successive Emperors. He furnished Kepler with an assistant and a printing press; and obtained for him the Professorship of Astronomy at the University of Rostock in Mecklenburg. Apparently, however, the Emperor could not induce Wallenstein to take over the responsibility of the 8000 crowns, still owing from the Imperial treasury on account of the Rudolphine Tables. Kepler made a last attempt to secure payment at Ratisbon, but his journey thither brought disappointment and fatigue and left him in such a condition that he rapidly succumbed to an attack of fever, dying in November, 1630, in his fifty-ninth year. His body was buried at Ratisbon, but the tombstone was destroyed during the war then raging. His daughter, Susanna, the wife of Jacob Bartsch, a physician who had helped Kepler with his Ephemeris, lost her husband soon after her father's death, and succeeded in obtaining part of Kepler's arrears of salary by threatening to keep Tycho's manuscripts, but her stepmother was left almost penniless with five young children. For their benefit Louis Kepler printed a "Dream of Lunar Astronomy," which first his father and then his brother-in-law had been preparing for publication at the time of their respective deaths. It is a curious mixture of saga and fairy tale with a little science in the way of astronomy studied from the moon, and cast in the form of a dream to overcome the practical difficulties of the hypothesis of visiting the moon. Other writings in large numbers were left unpublished. No attempt at a complete edition of Kepler's works was made for a long time. One was projected in 1714 by his biographer, Hantsch, but all that appeared was one volume of letters. After various learned bodies had declined to move in the matter the manuscripts were purchased for the Imperial Russian library. An edition was at length brought out at Frankfort by C. Frisch, in eight volumes, appearing at intervals from 1858-1870.

Kepler's fame does not rest upon his voluminous works. With his peculiar method of approaching problems there was bound to be an inordinate amount of chaff mixed with the grain, and he used no winnowing machine. His simplicity and transparent honesty induced him to include everything, in fact he seemed to glory in the number of false trails he laboriously followed. He was one who might be expected to find the proverbial "needle in a haystack," but unfortunately the needle was not always there. Delambre says, "Ardent, restless, burning to distinguish himself by his discoveries he attempted everything, and having once obtained a glimpse of one, no labour was too hard for him in following or verifying it. All his attempts had not the same success, and in fact that was impossible. Those which have failed seem to us only fanciful; those which have been more fortunate appear sublime.

When in search of that which really existed, he has sometimes found it; when he devoted himself to the pursuit of a chimera, he could not but fail, but even then he unfolded the same qualities, and that obstinate perseverance that must triumph over all difficulties but those which are insurmountable." Berry, in his "Short History of Astronomy," says "as one reads chapter after chapter without a lucid, still less a correct idea, it is impossible to refrain from regrets that the intelligence of Kepler should have been so wasted, and it is difficult not to suspect at times that some of the valuable results which lie embedded in this great mass of tedious speculation were arrived at by a mere accident. On the other hand it must not be forgotten that such accidents have a habit of happening only to great men, and that if Kepler loved to give reins to his imagination he was equally impressed with the necessity of scrupulously comparing speculative results with observed facts, and of surrendering without demur the most beloved of his fancies if it was unable to stand this test. If Kepler had burnt three-quarters of what he printed, we should in all probability have formed a higher opinion of his intellectual grasp and sobriety of judgment, but we should have lost to a great extent the impression of extraordinary enthusiasm and industry, and of almost unequalled intellectual honesty which we now get from a study of his works."

Professor Forbes is more enthusiastic. In his "History of Astronomy," he refers to Kepler as "the man whose place, as is generally agreed, would have been the most difficult to fill among all those who have contributed to the advance of astronomical knowledge," and again *à propos* of Kepler's great book, "it must be obvious that he had at that time some inkling of the meaning of his laws—universal gravitation. From that moment the idea of universal gravitation was in the air, and hints and guesses were thrown out by many; and in time the law of gravitation would doubtless have been discovered, though probably not by the work of one man, even if Newton had not lived. But, if Kepler had not lived, who else could have discovered his Laws?"

Appendix I

List of Dates.

1. Johann Kepler, born 1571;

2. school at Maulbronn, 1586;

3. University of Tübingen, 1589;

4. M.A. of Tübingen, 1591;

5. Professor at Gratz, 1594;

6. "Prodromus," with "Mysterium Cosmographicum," published 1596;

7. first marriage, 1597;

8. joins Tycho Brahe at Prague, 1600;

9. death of Tycho, 1601;

10. Kepler's optics, 1603;

11. Nova, 1604;

12. on Comets, 1607;

13. Commentary on Mars, including First and Second Laws, 1609;

14. Professor at Linz, 1612;

15. second marriage, 1613;

16. Third Law discovered, 1618;

17. Epitome of Copernican Astronomy, 1618-1621;

18. Rudolphine Tables published, 1627;

19. died, 1630.

Appendix II

Bibliography.

For a full account of the various systems of Kepler and his predecessors the reader cannot do better than consult the "History of the Planetary Systems, from Thales to Kepler," by Dr. J.L.E. Dreyer (Cambridge Univ. Press, 1906). The same author's "Tycho Brahe" gives a wealth of detail about that "Phœnix of Astronomers," as Kepler styles him. A great proportion of the literature relating to Kepler is German, but he has his place in the histories of astronomy, from Delambre and the more modern R. Wolfs "Geschichte" to those of A. Berry, "History of Astronomy" (University Extension Manuals, Murray, 1898), and Professor G. Forbes, "History of Astronomy" (History of Science Series, Watts, 1909).

Glossary

Apogee:

The point in the orbit of a celestial body when it is furthest from the earth.

Apse:

An extremity of the major axis of the orbit of a body; a body is at its greatest and least distances from the body about which it revolves, when at one or other apse.

Conjunction:

When a plane containing the earth's axis and passing through the centre of the sun also passes through that of the moon or a planet, at the same side of the earth, the moon or planet is in conjunction, or if on opposite sides of the earth, the moon or planet is in opposition. Mercury and Venus cannot be in opposition, but are in inferior or superior conjunction according as they are nearer or further than the sun.

Deferent:

In the epicyclic theory, uneven motion is represented by motion round a circle whose centre travels round another circle, the latter is called the deferent.

Ecliptic:

The plane of the earth's orbital motion about the sun, which cuts the heavens in a great circle. It is so called because obviously eclipses can only occur when the moon is also approximately in this plane, besides being in conjunction or opposition with the sun.

Epicycle:

A point moving on the circumference of a circle whose centre describes another circle, traces an epicycle with reference to the centre of the second circle.

Equant:

In Ptolemy's excentric theory, when a planet is describing a circle about a centre which is not the earth, in order to satisfy the convention that the motion must be uniform, a point was found about which the motion was apparently uniform,[4] and this point was called the equant.

> **Footnote 4:** I.e. the *angular* motion about the equant was uniform.

Equinox:

When the sun is in the plane of the earth's equator the lengths of day and night are equal. This happens twice a year, and the times when the sun passes the equator are called the vernal or spring equinox and the autumnal equinox respectively.

Evection:

The second inequality of the moon, which vanishes at new and full moon and is a maximum at first and last quarter.

Excentric:

As an alternative to epicycles, planets whose motion round the earth was not uniform could be represented as moving round a point some distance from the earth called the excentric.

Geocentric:

Referred to the centre of the earth; e.g. Ptolemy's theory.

Heliocentric:

Referred to the centre of the sun; e.g. the theory commonly called Copernican.

Inequality:

The difference between the actual position of a planet and its theoretical position on the hypothesis of uniform circular motion.

Node:

The points where the orbit of the moon or a planet intersect the plane of the ecliptic. The ascending node is the one when the planet is moving northwards, and the line of intersection of the orbital plane with the ecliptic is the line of nodes.

Occultation:

Usually means when a planet or star is hidden by the moon, but it also includes "occultation" of a star by a planet or of a satellite by a planet or of one planet by another.

Opposition

v. Conjunction.

Parallax:

The error introduced by observing from some point other than that required in theory, e.g. in geocentric places because the observations are made from the surface of the earth instead of the centre, or in heliocentric places because observations are made from the earth and not from the sun.

Perigee:

The point in the orbit of a celestial body when it is nearest to the earth.

Precession:

Owing to the slow motion of the earth's pole around the pole of the ecliptic, the equator cuts the ecliptic a little earlier every year, so that the equinox each year slightly precedes, with reference to the stars, that of the previous year.